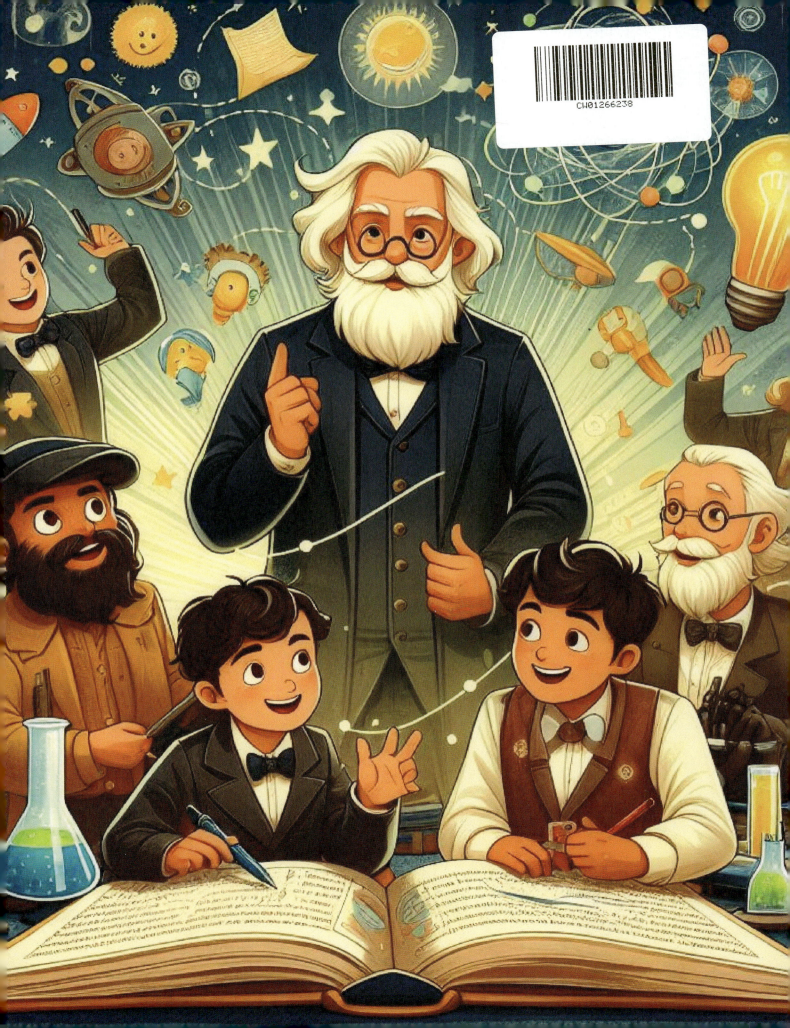

Once up on a time, there were a group of adventurous kids who loved to gaze up at the stars and wonder about the universe. They called themselves the "Cosmic Explorers," and every night, they would meet at the hilltop observatory, their eyes wide with excitement and heads filled with dreams of discovering the unknown would read their book about the universe.

One clear evening, as the stars began to twinkle, the eldest of the group, Ellie, opened an old book that she had borrowed from her school library that day called "Great minds of science".

As the stars above began to twinkle, the group huddled around Ellie, their eyes wide with anticipation.

"Let's embark on a journey through the minds of those who've unraveled the secrets of the universe," Ellie announced, opening to the first chapter.

1. Isaac Newton (1642-1727):

Imagine if apples had superpowers. Well, one apple helped Sir Isaac Newton discover gravity! He saw an apple fall from a tree and started to wonder why it fell straight down, instead of sideways or up.

This led him to come up with the law of gravity, explaining not just why objects fall towards the Earth but also how planets orbit the sun.

2. Albert Einstein (1879-1955)

Turning the page, Ellie continued, "Next, meet Albert Einstein, a superhero of science, who explored time and space."

He came up with the theory of relativity, which sounds complicated but is about understanding how space, time, and gravity work together.

One of his famous ideas is that the faster you travel, the slower time goes, which is a bit like imagining if you could run so fast, you'd see a clock ticking slower.

3. Marie Curie (1867-1934):

As the moon climbed higher, Ellie introduced Marie Curie, the relentless scientist who discovered radioactive elements.

She was like a treasure hunter, but instead of searching for gold, she discovered two new elements: polonium and radium. She was the first woman to win a Nobel Prize and the only person to win Nobel Prizes in two different sciences (Physics and Chemistry)! Her work helped doctors use x-rays to see inside our bodies.

4. Galileo Galilei (1564-1642):

Next, Ellie brought them closer to the stars with Galileo Galilei, the man who brought the stars closer to us.

He improved the telescope, which is like a big spyglass, and used it to discover mountains on the moon, moons around Jupiter, and more. He showed us that the Earth and other planets go around the Sun, not the other way around, which changed how we see our place in the universe.

5. Charles Darwin (1809-1882):

Ellie's storytelling then sailed across the oceans with Charles Darwin on his voyage that led to the theory of evolution.

Imagine going on a long boat trip and coming back with an idea that changes the world. That's what happened to Charles Darwin. He developed the theory of evolution, explaining how all living things change over time and how we're all connected in the big family tree of life.

6. Stephen Hawking (1942-2018):

Under the glow of the moon, Ellie spoke softly of Stephen Hawking.

"Despite being in a wheelchair and not being able to speak without help, explored the secrets of the universe, from black holes to the big bang!" "That's really brave," whispered Josh, looking up at the sky. "Exactly!" Ellie reassured, he showed us that the universe has more wonders than we can imagine.

7. Jane Goodall (1934-present):

"Next in line, is Jane Goodall" Ellie said turning the page to the next chapter. "Imagine living in the jungle to make friends with chimpanzees. That's what she did!"

"She showed us how much we can learn by just watching and listening to animals, especially chimps, who are a lot like us. Her work helps protect animals and their homes."

"I want to be like her," declared Emma, with a nod.

8. Nikola Tesla (1856-1943):

"Tesla was a wizard with electricity," Ellie said, sparking the kids' imaginations with tales of lightning and wireless energy. He invented ways to send electricity through the air and made the first alternating current (AC) motor, which helps power everything from homes to schools without needing wires everywhere.

"Hey, I saw the Tesla coil at the museum that I visited with my dad!" Emma exclaimed excitingly.

9. Rosalind Franklin (1920-1958):

"Rosalind Franklin was like a detective with a very special camera." Ellie continued, "She took critical photographs of DNA that helped other scientists figure out it's shaped like a twisted ladder, called a double helix. Her work was key to understanding how traits are passed from parents to children."

"Like a 'recipe book' for building any living thing?" asked Joey, intrigued. Ellie laughed with at the imagination and said, "Yes, you can call it that."

10. Thomas Edison (1847-1931):

"And then there's Thomas Edison, was an inventor who turned his ideas into things we use every day. He lit up the night with his invention of the light bulb," Ellie added, gesturing to the glowing lamps around them.

"He made it so we could stay up late reading and exploring! "

11. Carl Sagan (1934-1996):

Ellie's face lit up with wonder as she talked about Carl Sagan, the storyteller of the stars. "He was not only a brilliant scientist who studied planets and stars but also a storyteller who helped people feel like they were part of the big, mysterious universe"

Sagan taught us about the importance of science in our lives and inspired us to wonder about what's out there, beyond our world. He believed that by exploring space, we learn more about ourselves and how to take care of our home planet.

12. Neil Armstrong (1930-2012):

For the final tale of the night, Ellie recounted Neil Armstrong's lunar adventure. "He was an astrophysicist who took the very first human steps on the moon. Just think, one of you might step on Mars!"

His famous words, "That's one small step for man, one giant leap for mankind," remind us of the human drive to explore and discover.

With each chapter, the Cosmic Explorers felt more connected to these legends of science. As they promised to keep looking up and wondering, they knew their adventure was just beginning, under the same stars that had guided those great minds before them.

The End

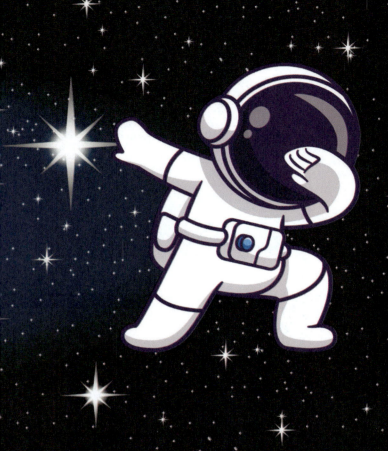
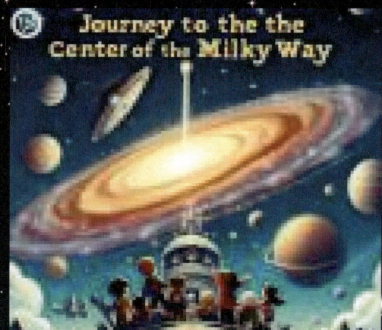

I hope you enjoyed this little story.

If you are interested in more, check out my other books on different planets in the solar System or even take a journey on a Space Ship!

Printed in Great Britain
by Amazon